GEODES

Timothy Cunninghamm

Running Press • Philadelphia, Pennsylvania

Copyright © 1992 by Running Press

All rights reserved under the Pan-American and
International Copyright Conventions.

This book may not be reproduced in whole or in part in any form or by any means, electronic or mechanical, including photocopying, recording, or by any information storage and retrieval system now known or hereafter invented, without written permission from the publisher.

Canadian representatives: General Publishing Co., Ltd., 30 Lesmill Road, Don Mills, Ontario M3B 2T6.

International representatives: Worldwide Media Services, Inc., 30 Montgomery Street, Jersey City, New Jersey, 07302.

9 8 7 6 5 4 3 2
Digit on the right indicates the number of this printing.

Library of Congress Cataloging-in-publication Number 92-53690

ISBN 1-56138-144-6

Edited by Cynthia L. Gitter
Package and cover design by Toby Schmidt
Book interior design by Christian T. Benton

Package front geode photographs: Courtesy Department of Library Services, American Museum of Natural History; ruby geode photograph by Steve Belkowitz Photography.

Book cover and package back photographs: Top row (Jeff Scovil): Quartz geode (90-42-4); Millerite and pyrite geode (90-43-31); Millerite and pyrite geode (90-44-2). Middle row (Jeff Scovil): Laguna agate (92-9-2); Apache agate (92-9-6); Laguna agates (92-9-11). Bottom row: Laguna agate (92-9-16), Jeff Scovil; Laguna agate (92-10-3), Jeff Scovil; Dugway geode, Irving Fisher. Package back, bottom: split geode photograph by American Museum of Natural History.

Package interior background: photomicrograph by Irving S. Fisher.
Additional photography by Steve Belkowitz Photography.

Book interior photographs: American Museum of Natural History: p. 8 (J. Beckett/R. Raeihle), p. 13 (Beckett/Perkins), pp. 23, 30 (#3629). Badlands National Monument/Grant, p. 71. Free Library of Philadelphia, p. 10, top. Japan Airlines, p. 18. Barbara Lawlor, p. 9. National Park Service/Parker Hamilton (#115-1922), p. 56. Oregon Dept. of Geology and Mineral Industries: p. 11 (Leo F. Simon), p. 12, top (Lewis Birdsall), p. 40. Oregon Dept. of Transportation, p. 22 (#8333). Photo Researchers: p. 12, bottom (© 1986 Wetmore), p. 32 (#1327 © Josephus Daniels), p. 46 (#7P9745 © Renee Lynn). PA Dept. of Transportation, p. 48. Jeff Scovil: pp. 14, 33, 51, 59. H. Stern: pp. 15, 53. U.S. Geological Survey: p. 10, bottom (Lee, W.T. 3307), p. 20 (Wentworth, C.K. 632), p. 24 (Russell, I.C. 725), p. 27 (King, P.B. 460), p. 29 (Richter, D.H. 75)

Book interior illustrations by Helen I. Driggs

Typography: Gill Sans, by Richard Conklin, Philadelphia, Pennsylvania

Printed in the United States by The Collett Company

The author would like to extend his special thanks to Robert Meyers, Bria, The Rock Rollers Club of Spokane: Ed, Jim, Harland, and Ira J.

Running Press Book Publishers
125 South Twenty-second Street
Philadelphia, Pennsylvania 19103

Timothy Cunninghamm received his first patent in 1991 for an optical sundial. He teaches science classes and brings the wonders of science to children of all ages through the dr. Kno Science Shows.
He is currently developing interactive exhibits for the Inland Empire Discovery Center in Spokane, Washington.

CONTENTS

Introduction
5

1. What Are Geodes?
7

2. The Mystery of Geodes
What Are Geodes Made From?

The Recipe for a Perfect Geode

17

3. In Search of Geodes
Before You Explore

Bringing Home a Bunch

Keeping Your Journal

Looking for Likely Spots

Great Geode Discoveries

35

4. How to Make Your Own Natural-Looking Geodes
Making the Geode Shells

Growing Ruby-Colored Geodes

61

Introduction

Geodes are riddles inside a shell.

What are these mysterious rocks that beckon us from the depths of the earth? What are these crystal flowers that grow so peacefully in stone, slumbering beneath the earth? What stories do they tell?

Geodes are jewels wrapped inside rocks—colorful, magical crystals, dazzling in the light, that have been covered for thousands of years. Do these natural treasures formed by the warmth of the earth hold clues about the formation of the universe?

Philosophers and poets have long pondered the magic of geodes. How do crystals grow inside rocks buried in the earth? Do they contain treasures, or could they be omens of things to come?

Sought after by magicians for what they believed was healing power, geodes have fascinated seekers around the world. To discover and open up a geode invites the viewer to a magical world never before seen.

ONE

What Are Geodes?

GEODES

What are geodes? How old are they? How are they formed? Where can I find one?

The word *geode* means "like earth." Geodes—crystals encased in rock—are a family of rock formations that include thundereggs and agates. These three formations are related to each other and are found in similar environments, but they have characteristics that make each of them unique.

The outside of most geodes looks like the surface of a small rocky planet, or a bumpy mud ball. The hollow space inside a real geode is lined with mineral deposits. These colorful bands of crystallized minerals are layered from the inside of the outer shell toward the center. In the middle are larger crystals that point to the hollow open center. Often these spiky crystals are colored the hues of various minerals. The variety of colors in geodes is endless. Sometimes the hollow section is filled with water that could be a million years old. In Australia, oil-filled geodes have been found. No two are ever alike.

Amethyst geodes

Most geodes are the size of a baseball, but some are as small as a marble! On the other hand, some rare

What Are Geodes?

forms of geodes are so large that you could crawl inside. Geodes have been a part of folklore and mythology for centuries. The Legend of King Arthur and the Knights of the Round Table refers to the crystal cave of Merlin the Magician. A secret place of rest and meditation, this inner sanctum may have been a large geode. It was believed that Merlin's magical powers were enhanced by

This geode would be an impressive addition to anyone's collection.

spending time inside this powerful cave.

According to the legend, Merlin may have been inspired by this mystical place to dream up the construction and engineering of the mysterious arrangement of huge stones in England known as Stonehenge.

Thundereggs are spherical or odd-shaped rocks that contain the same

Did Merlin draw his magical powers from a giant geode?

Just below the surface of the earth lies a world of magic. This is Mirror Lake, Grand Caverns, Virginia.

What Are Geodes?

This thunderegg, uncovered in Oregon, has a star hidden inside!

colorful banded minerals that you might find in a geode. These bands are a type of quartz called chalcedony (kal-SED-onee). Some thundereggs have been found to contain agate, jasper, and even opal. The crystals in a thunderegg are more dense and of a much finer grade than in most geodes. The layers of colors are just as brilliant and varied, and the crystals often form geometric patterns not found in geodes. Thundereggs are found all over the world, but the most famous deposit is in Oregon.

GEODES

From the mythology of the Warm Spring Indians comes the story of how the Thunder Beings in the Cascade Mountain Range played ball with eggs stolen from the nests of thunderbirds. In lofty places high upon Mt. Hood and Mt. Jefferson, the Thunder Beings played a violent game of catch. When the eggs collided with each other, they created sparks and heat, or what we know as lightning, while thunder came from the crash of the eggs hitting each other. Minerals crystallized from the intense heat created at the moment of impact and formed a picture of places and things to come. The thundereggs fell to earth, where they were collected and revered as a source of healing and prophecy. Thundereggs were often split apart by heating them beside a fire and then splashing them with cold water.

Agates are made from the same type of minerals

Digging for thundereggs on a ranch in Oregon

Thundereggs in the making? No, just lightning over the mountains.

What Are Geodes?

It's easy to see how bull's-eye agate got its name. This specimen is from Brazil.

as geodes and are formed much like geodes and thundereggs. In fact, an agate is like a thunderegg with little or no shell. Weathering and erosion have worn away most or all of the shell.

Agates are formed from a finer grade of crystal than geodes. The crystals are banded into layers of many colors determined by the minerals present when the agate was formed.

Agates form in the bubbles and empty spaces in rocks. It takes thousands and sometimes millions of years for a piece of agate to grow. As gem-quality stones, agates have been made into decorative artwork and jewelry for thousands upon thousands of years.

The Old Testament mentions the use of agate as one of the stones in the breastplate of the Hebrew priest Aaron, a representative of the tribes of Israel. You might recognize some agate at the drugstore in the form of an antique mortar and pestle once used to grind minerals and chemicals.

Seals were once used in place of signatures on official documents. This seal is carved from carnelian, a reddish mineral of quartz.

A rock hound is a person who collects rocks as a hobby. Rock hounds love agates because of their wonderful colors and because no two pieces are alike.

Agates are also favored by lapidary artists, or people who cut precious stones, because they are soft enough to shape into jewelry, and yet durable enough to be worn without breaking.

Geodes, thundereggs, and agates all contain mysterious patterns that dazzle the eye and are of a unique design. They have been collected and prized as treasures

What Are Geodes?

for thousands of years—and yet there are still many unanswered questions as to how they were formed.

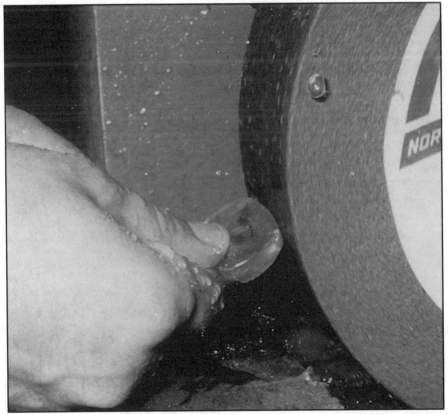

The grindstone shapes rough stones into beautiful gems—but watch your thumbs!

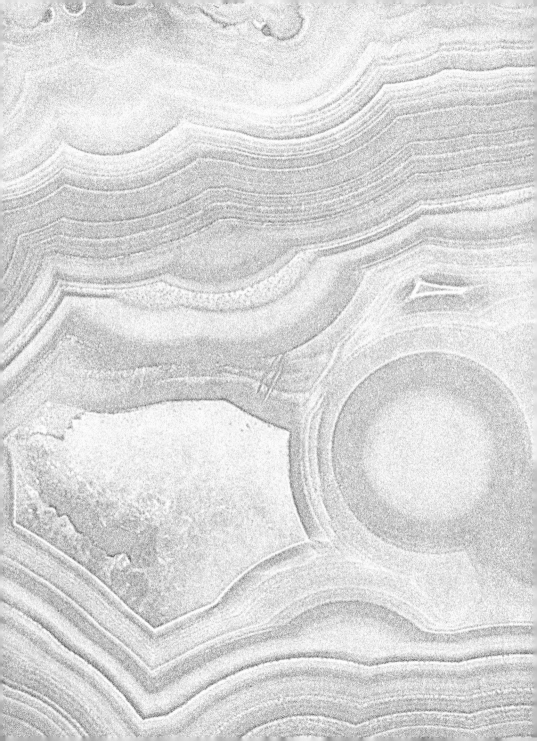

TWO

The Mystery of Geodes

GEODES

The exploration of earthly mysteries, such as the origins of geodes, belongs to the school of geology. *Geo* means earth, and *ology* means to discuss or study.

We live on the cooled surface of a sphere containing metal and hot liquid rock. This round ball that we call earth has a super-hot liquid center with a cooled surface crust. This crust, which we call the ground, is actually a thin layer made of three types of rock. These three types are igneous, sedimentary, and metamorphic rock. The dirt at your feet, the mountains in the distance, and the shore of your favorite beach all come from these three types of rock.

The earth is letting off steam—and perhaps cooking up some geodes—near this Japanese volcano.

The Mystery of Geodes

Igneous means formed by fire. Like a volcano, the interior of the earth is very hot. At the very core of the earth are the elements that form the liquid center of this planet. This molten rock, or magma, is under great pressure. The molten rock finds its way to the surface of the earth, often through volcanoes. Along the way this liquid rock picks up various minerals, cools, and becomes solid igneous rock.

Igneous rock makes up 95% of the earth's surface. Granite, basalt, and rhyolite are examples of igneous rock. As magma flows to the surface and cools, it creates bubble holes that can collect minerals and water. As these rocks cool even more, the water evaporates and leaves small pockets of crystallized minerals. This is one of the many ways that geodes are created.

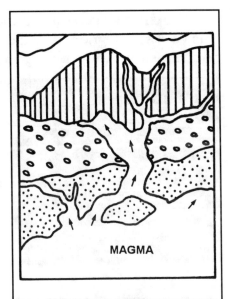

GEODES

Sedimentary rocks are the children of igneous rocks. As igneous rocks wear down from wind, rain, and erosion, they break into smaller and smaller particles. The igneous rock begins to look more like sand. The sand particles and small rock pieces are blown about by the wind and carried by streams until they come to rest in valleys or at the bottom of lakes.

Stone sculpted by flowing water

Layer by layer the sandy deposits build up, together with matter from dead

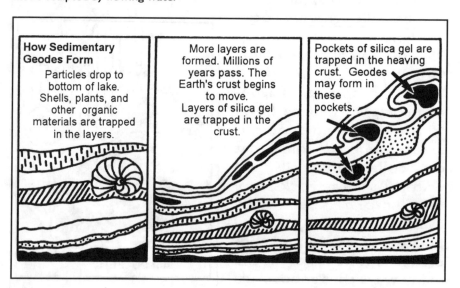

How Sedimentary Geodes Form
Particles drop to bottom of lake. Shells, plants, and other organic materials are trapped in the layers.

More layers are formed. Millions of years pass. The Earth's crust begins to move. Layers of silica gel are trapped in the crust.

Pockets of silica gel are trapped in the heaving crust. Geodes may form in these pockets.

The Mystery of Geodes

Devil's Tower, Wyoming, was formed by lava pushing out of the earth.

trees, leaves, and fish and animal bones. The combination of rock and organic matter forms in layers.

As the organic material decays, the bones are sandwiched and preserved between the layers of sand and clay. Each layer adds pressure to the ones formed before it. Sedimentary layers are where most fossils are found. Sometimes the pressure becomes so great that rocks can be bent and twisted like a ribbon. The movement of the earth can cause these ribbon layers to fold on top of one another. When this happens, a tiny piece of mineral or crystal can get caught inside a gas bubble. This bubble grows a hardened shell that can become a sedimentary geode or thunderegg.

GEODES

These sedimentary limestone "chandeliers" hang from the ceiling at Oregon Caves National Monument. They formed one drop at a time over thousands of years.

Metamorphic rock is the third kind of rock in the earth's surface. The word *metamorphic* comes from the Greek word that means to change, and this type of rock is formed from the transformation of the other two types of rock. Deep in the earth, heat and pressure squeeze layers of igneous and sedimentary rock together. The intense heat melts the rock and re-forms it into a new mixture.

Sandstone is turned into quartzite, and limestone becomes marble. The valuable ores of gold and silver, platinum, diamonds, and all other rare earth minerals have metamorphic beginnings. Metamorphic rock is the only type in which geodes do not form.

Thanks to the movement of the earth, you don't always have to dig to find these treasures. Over millions of years some of these deposits have been pushed up to the surface. A great many lake beds, riverbanks, and desert floors are just waiting to be explored.

What Are Geodes Made From?

All geodes—like all rocks—are made from minerals. One of the most abundant minerals found in the earth's crust is *quartz*, the main ingredient of every geode.

During the last 50 years, the use of quartz in inventions has grown

Quartz crystals—the building blocks of all geodes

dramatically, from computers and optics to communications and satellites. In your home today, nearly every electrically powered appliance contains some form of quartz—the same mineral found in all geodes. Without quartz we wouldn't have geodes.

Earth on the move. What secrets will be uncovered?

The Recipe for a Perfect Geode

You may not feel it most of the time, but the crust of our earth is moving. When these movements are too quick they can be dangerous, such as an earthquake or volcano. Most movements take place over a long period of time and are too slow for us to notice. Millions of

The Mystery of Geodes

years go by as the earth pushes up layers of rock, creates mountain ranges and forms the valleys that will become oceans. You may be standing or sitting on the bottom of what was once a great ocean. Geodes got their start here, too!

Let's take a peek into the earth's kitchen and see what it takes to whip up a geode.

Another Favorite From Mother Earth's Kitchen

A Recipe For: SEDIMENTARY GEODES

1 HANDFUL SILICON DIOXIDE (QUARTZ MAY BE SUBSTITUTED.)

2 CUPS WATER

A PINCH OF CLAY

1 PART IRON ORE

DASH OF TRACE MINERALS (FOR COLOR)

HEAT AND MIX WELL. ADD TO DEPRESSION IN THE OCEAN FLOOR.

SET TIMER FOR A MILLION YEARS.

Let's take a look at what happens over that million years:

A gel-like mixture of mostly water and silica (silicon and oxygen mixed) settles on the ocean floor and is pushed under the surface of the earth. Layers of sediment build up around the little ball of gel. As the nodule moves deeper, the weight of the sediment creates pressure, and as the pressure rises, the temperature goes up. The heat causes the silica to harden into a shell, and soon the shell is thick enough to stand up to the many layers of sediment building up on the top and sides. The pressure and the heat eases for a while—maybe a few hundred years. The outer shell begins to harden, but the inside is still soft and gel-like.

Soon more layers of sediment are added on top. The heat begins to climb once more, and this time the gel inside starts to stick to the inner wall. This first layer is mostly silica in the form of quartz crystal. Soon the next layer begins to stick to the newly formed inner layer of quartz. This material is finer than the coarse quartz that formed the outer shell.

Small amounts of other minerals start to show their color: the iron produces specks of red, traces of nickel and chromium add some green to the next few layers.

Thousands of years fly by as the silica gel separates into layers of crystallized minerals. Each new layer is a different color from the previous one.

Soon the shell has to release some heat and pressure. A small vent appears on the side of the geode. As the watery silica gel is squeezed out, the shell shrinks just a little bit under the incredible pressure. The gel leaves a trail of color from the center to the outer shell.

The geode is buried deeper and deeper into the earth. The pressure and temperature are climbing! Soon the water that was a part of the silica gel is just about evaporated. What's left is a succession of finer and finer layers of crystal growing toward the center of the geode.

Slowly the temperature begins to drop. The earth's oven returns the geode to the surface. This is a slow journey, because the geode is close to 10,000 feet below the surface of the earth.

There is now only a trace of water and silica left in the nodule. The lower heat and pressure allow larger crystals to form. Slowly these larger crystals begin to grow. Over the next half a million years, they produce a unique, original formation.

Let's try a variation on this recipe. This time, let's get out of the water and into the fire.

The stripes in this sedimentary rock show how it was built up layer by layer.

> **Another Favorite From Mother Earth's Kitchen**
>
> **A Recipe For:** IGNEOUS GEODES
>
> POUR THE SAME INGREDIENTS AS BEFORE INTO A BUBBLE OR CAVITY IN A HOT LAVA FLOW.
> HEAT AND MIX WELL.
> SET TIMER FOR A MILLION YEARS.

    As the silica gel hits the molten rock cavity, the water turns to steam. Exposed to the super-heated rock, the silica changes instantly into a hardened glass-like shell. The remaining gel is trapped inside the shell. The cavity, or host rock, is really hot, so activity is sped up a bit.

    Even though the outer wall has fused into a semi-hard surface, the gel on the inside is still flexible. Layer by layer, this gel begins to separate. Each layer adds its own color signature and follows the inside contours of the geode. This layering continues as conditions outside the shell undergo several changes.

    Thousands of years pass by, and the earth undergoes

The Mystery of Geodes

From such fiery beginnings come some really cool geodes.

some important changes. When the geode started its journey, the lava flow was on the earth's surface. Several lava flows have covered up the geode since then, so the only way for the geode to resurface is for wind or water to wear away the rock that has buried it. This will take thousands and thousands of years, but finally the geode will be freed from its parent rock.

Keeping its round shape, this newly formed rock will roll into a stream bed and be swept underground for another ten thousand years or so. As it is pushed down and reheated, there will be even more changes. The interior colors will become brighter and deeper, and the geometric patterns on the inside will be pushed out toward the edges of the hardened shell. Upon resurfacing, the outer shell may become smoother.

Chef's Notes

In the sedimentary geode recipe, when the cooling stages begin, the core of the geode shrinks. The force acting upon the geode can be great enough to draw in new minerals that create little pathways of sedimentary material into the geode. The shrinkage can also cause the crystalline structures to pull away from the inside of the shell, creating a star-like pattern. The thundereggs of Oregon show this pattern.

A sedimentary thunderegg from Oregon

> **Another Favorite From Mother Earth's Kitchen**
>
> **A Recipe For:** _AGATES_
>
> POUR THE SAME INGREDIENTS AS BEFORE INTO A CAVITY IN A ROCK, GEODE, OR PIECE OF WOOD.
> HEAT AND MIX WELL.
>
> SET TIMER FOR A MILLION YEARS.

This recipe is getting familiar, isn't it? Surprisingly enough, the agate is the most famous of the nodular—or round, lump-like—rocks, perhaps because the agate is a master of disguise.

Agates can form by intrusion, which means that the silica gel can work its way into a cavity in a rock, geode, or hollow piece of wood. If it follows the heat-and-pressure process, it will end up as an irregularly shaped agate. Weathering and erosion free the agate from the host rock, leaving bits of rock clinging to the outside. Any geode-type nodules might contain some agate, because it can force its way into them while they're forming. When the geodes are cut open, you might find a little agate mixed in with the quartz crystals.

Agates can also be formed when minerals and silica are deposited into an igneous rock bubble or opening. In Oregon, bands of ledge agate have been found, measuring one foot thick and 30 feet long, that formed in this way between layers of rock.

The agate recipe is the same as for the igneous geode. After millions of years of incubation underneath the earth, the weatherizing process frees the agate from the host rock.

Deposits of agates left inside the host rock can be easily seen sticking out from the igneous formation. Since they are still in the same situation as when they were formed, they are known as agates "in situ."

An agate's surface can be rough as a rock in places, with only glimpses of the shiny interior. Some agates

As water races, it wears away the surface rock, sometimes exposing geodes.

The Mystery of Geodes

have a thin covering of dirt or bits of rock, while others can be completely uncovered. An uncovered agate has a dusty surface and is well disguised in surrounding rock.

Another way to form an agate is to wrap a small crystal or fossil inside a gel-like mixture of silica. Add many layers of sediment around this little ball, and the pressure and heat begin to build. If a silica mixture follows the same process as in the other recipes, in a million years or so it could become a gem-quality agate.

The colorful banding that occurs in agates is created in the same way that it is formed inside the geode.

Chef's Notes

The formation of the geodes, thundereggs, and agates are closely related. The presence of certain trace minerals will give you a clue to their original location. Most rock hounds can identify the location of a geode by the color and shape of the internal banding.

Now that we know more about the mysterious geode, it's time to get our hands dirty. Let's go exploring.

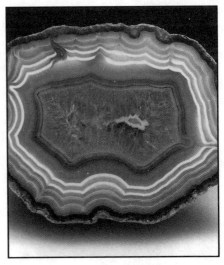

An agate geode from Chihuahua, Mexico

THREE

In Search of Geodes

GEODES

You might want to start your geode search in your back yard or an empty field near your home. Or you could begin by going to the local library to look for information that might lead you to some better locations in your area. Several books and magazines include maps and directions to start you off. For instance, United States Geological Survey produces maps and regional guides to areas of interest to rock hounds. You can find these maps at your local map store. Or you might let your fingers do some searching in the phone book!

There might also be several rock-hunting clubs in your city. The best way to find these clubs is to look in the phone book under Lapidaries. Many clubs operate out of hobby shops where rock hounds buy their equipment. In addition to carrying tools and books, these shops are full of samples of rocks and minerals. You can usually find quite a collection of geodes, thundereggs, and agates amid the many geological treasures. Better still, you may find the person who collected these specimens! He or she will probably be happy to answer any questions you have. Ask to see some geodes. Show your interest in collecting some of the treasures in the area. These expert rock hounds are likely to tell you tales of their own rock-hunting trips. They might even tell you where their favorite spots are!

If your new friends are members of the local rock-hunters club, they'll try to get you to come to a meeting. All rock hounds love to show and tell. And hearing first-hand experience from a true rock hound is a great way to kick off your own search.

Experienced rock hounds are fond of sharing their knowledge, especially if they think you are a budding "pebble pup." Don't be embarrassed if they call you that. It's an affectionate nickname for a young rock hound.

Most lapidary stores carry several small field books that will help you scout out your local area. Some of these books are short, to the point, and discuss only the types of rocks or gems found in your city or county. If you're the independent type who would like to strike out on your own, a book like this will help you figure out what kinds of rocks are associated with the treasure that you are seeking.

Before you go running off to the nearest field, though, take some tips from some of the wise rock hounds who have been doing this for many years. The safety measures you take will guarantee you'll be around to tell many of your own tales!

Before You Explore

All experienced rock hounds agree on these important rules:
1) Always tell an adult where you're going.
2) Always leave directions to the location you'll be exploring.

3) Always take a partner. (You'll need help carrying all your treasures back home anyway.)
4) Always get permission to explore on private property.
5) Never, never explore an old mine shaft, well, or cave. These areas require special equipment.
6) Read tip #5 again. This means you!

Bringing Home a Bunch.

Before you set out to become a famous prospector or rock hound, you'll need to assemble your rock hunting kit. Here are some tools you'll want to bring along:

MY ROCK HOUND FIELD KIT CHECKLIST

- ☐ SMALL PICK OR HAMMER
- ☐ ROCK CHISEL
- ☐ SAFETY GLASSES
- ☐ POCKET KNIFE
- ☐ BANDAGES AND ANTISEPTIC
- ☐ JOURNAL AND PEN
- ☐ COMPASS & MAPS
- ☐ MAGNIFYING GLASS
- ☐ SPECIMEN BAG
- ☐ GLOVES
- ☐ ROCK IDENTIFICATION HANDBOOK

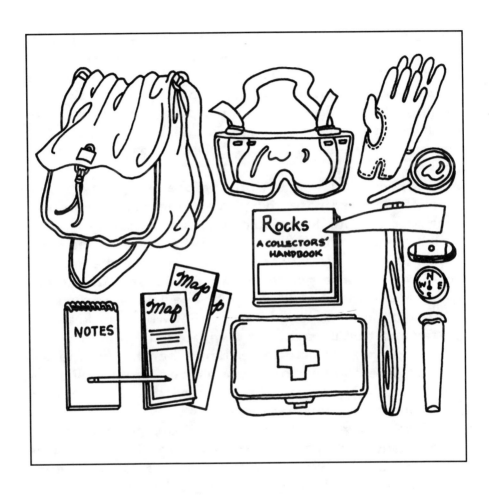

Let's go through the list. A good rock hound always carries a small pick or hammer for breaking stones away from the rock in which they are lodged. You can even get a combination pick hammer that's built for this purpose. You should find one that weighs about a pound. Get a clip for your belt to hang your hammer on. You'll need your hands free for climbing and scrambling around. There are also picks with hoes on the other side for digging. If you don't have one of these handy tools available, take along a small hammer and a rock chisel. Start small. Then add to your collection as your interest grows.

Always read instructions before using any equipment, and wear protective gloves. Another very important part of your kit should be a pair of safety glasses. Striking rocks with a hammer produces fast-flying chips which can be very dangerous! Don't take a chance with your sight. Some glasses strap on so you don't lose them, and some look like regular eyeglasses. Some even come with fancy colored frames. Get whatever kind you prefer and don't leave home without them!

A compass is also important when hiking and exploring. The compass does double duty when collecting rocks because you can use it to tell directions and to test your specimens for magnetism.

Take along some antiseptic and a few bandages. Some rocks can have sharp edges, and it's best to be prepared in case you get a few minor cuts or scrapes.

Find a small hand lens or an old magnifying glass. Some specimens deserve a close-up look.

You'll definitely need a specimen bag. An old canvas bag works well, but you can use anything that's handy and can stand up to the abuse of rocks rolling around in it. Be sure to use something that has straps or a good

In Search of Geodes

Geode-hunters' tools and jewels

handle. One of the best specimen bags is made from an old backpack. You'll need the larger space for specimens and tools, and the smaller compartments can hold your smaller tools and your journal.

In your kit be sure to include some maps and a handbook from the library or lapidary store. You can also purchase a mineral guide that has color photos of minerals and rocks. These guides are great for identifying specimens in the field.

Keeping Your Journal

Write it down! The best tool for a short memory is a sharp pencil. Make a journal to record your discoveries. A small three-ring notebook will do. The best treasure maps are the notes that others have handed down through the years. Keep the tradition alive!

Record your finds and locations. Using a rock identification manual, include the date of each find, color, hardness, species, and common name of the specimen.

JOURNAL 2   PAGE 26
SPECIMEN NUMBER _____ (Is this the first find of today?)
COMMON NAME _____ (If you can't identify the rock, describe it instead.)
COLOR _____ (Or colors)
HARDNESS _____ (Follow the Mohs scale)
LOCATION _____ (Where did you find it?)
DATE _____ (Date of find)
REMARKS _____   _____   (Write down anything unusual about your specimen. Was it hard to find? Leave yourself some clues about your hunt.)

Use your pocketknife to test your specimens for hardness. Copy the Mohs scale of hardness on the next page into your journal.

Mohs Scale of Hardness

Use this scale to test your minerals for hardness. #1 is the softest. You can scratch talc with your fingernail. #2 is a little harder, but you can still scratch it with your fingernail. You can't scratch calcite with your fingernail, but you *can* scratch it with a copper penny, and so on up the scale. Use this scale when identifying your specimen. After you have judged the color and shape, test the hardness and see if it fits the mineral profile in your handbook.

Softer → Harder		
	1) Talc	
	2) Gypsum	2½ = fingernail
	3) Calcite	3 = copper penny
	4) Fluorite	
	5) Apatite	5½ = pocketknife
	6) Orthoclase	6½ = steel file
	7) Quartz	
	8) Topaz	
	9) Corundum	
	10) Diamond	

Sample Field Labels

You'll want to keep track of your specimens, so attach labels to them before storing or displaying.

NAME _____ (of the rock) _____

DATE _____ (when found) _____

PAGE # (of your journal, where this rock is listed) _____

Keeping a journal might pay off someday. You may want to return to the exact spot where you found a beautiful gem-quality specimen. Or, while you're in the field, you may toss a rock in your bag for closer inspection at home. Later, at the local Rock Rollers Club, you may find out you've really discovered a new mineral. Luckily, you've written down exactly where you found that precious stone. You even thought to include a rough map to that spot. That's brilliant thinking—the mark of a true rock hound. You lead the club back to the exact spot, and now there's a mineral named after you. Believe it or not, it's happened before, so keep an up-to-date journal!

GEODES

A ranger helps search for geological clues in California.

Rock hunting is the best way to really see the earth up close! You will get your hands dirty along with your clothes, so take care to choose clothing that can get dirty and that will offer you some protection while walking, climbing, and looking under rocks. A pair of old leather gloves and a sturdy pair of shoes or hiking boots are good companions on a long walk.

Always keep an eye on the weather, too. A light jacket fits easily into your backpack and can protect you from the wind. Dress in layers. As the day warms up or cools down, you can add or remove clothing to stay comfortable.

Before you set out for a day of exploring, be sure to go through your checklist. Follow the rules for safety, then hit the trail and have fun.

Looking for Likely Spots

Geodes, thundereggs, and agates can all be found in the same areas. Sedimentary layers contain a wide range of rock types, so this is the best place to start. Look for areas that have been exposed by the weather or construction.

While riding down the highway around your city, look at the sides of the road. Have the roads been cut into the earth? Are there any hilly areas nearby? Look for the exposed side of a hill that has little or no vegetation. Are there alternating layers of different-colored rock, clay, or mud? Look closely, because these layers might look only slightly different. Look for white limestone deposits, or layers of volcanic rock. Is there an ash layer showing? Is there an eroded gully or ditch at the bottom of the hillside? Runoff from rainwater will uncover small and large rocks and expose even more layers, so this might be a good place to explore.

Keep looking. Go back to the same location several times. Every winter and every storm may uncover something new.

Another good place to go prospecting for treasures is around lakes or rivers in your area. Sedimentary

Road cuts can tell you much about the geology of your area.

geodes form along lake beds. Look at the banks of the river. Are there any layers, or strata, showing? Can you see different kinds of deposits? Remember that the latest deposits are on top, and the oldest are on the bottom. Look closely. Geodes may be hiding here. Watch the inside corner of each bend in the river. This is a good place to look for deposited rocks and agates.

If you hunt geodes at roadside, be sure to keep a safe distance from the highway.

Always be careful around rivers and streams. The ground is usually more slippery than you think! If you want to avoid the water completely, you might want to try gravel beds, dry lake beds, and old dry stream beds, which make excellent spots for exploring and rock collecting.

Agates can be found near igneous rock formations. A large, flat valley surrounded by mountains is a likely location to explore. There should be good runoff from rains and perhaps some gullies or eroded channels nearby.

Look for the striated, or striped, layers of sedimentary deposits. Keep an eye out for gravel beds and grassy areas next to large rock beds. Remember that new rocks are exposed every year, so you can always find new specimens. Check to see if the area has had any volcanic activity, even if it happened thousands of years ago. Layers of ash are great places to hunt for the elusive agate!

The best tool you can have is a sharp eye. What looks like an odd piece of gravel might really be an agate. Look closely. And don't worry—as time goes by you'll be collecting with the best of them.

You'll find your share of "leaverites," too! Everybody who's new brings home a bunch. They're not dangerous. It's just that when you show one to experienced rock hounds, they'll tell you to "leaverite" where you found it.

Great Geode Discoveries

Once you've brought your rocks home, what do you do next? If you've found a geode, you'll need some help opening it. Do not try to shatter the rock—you could hurt yourself. Your local lapidary shop has a saw that is made just for this job. They can also show you how to cut and polish your specimens. Agates can be tumbled, as can many of the stones you find. Tumbling removes the rocky covering, smoothing out the rough edges and polishing your treasure to keep and display.

When geodes, thundereggs, and agates are sliced open, they reveal intricate patterns of chalcedony, agate, or other crystal structures. As with most treasures, you won't know what you've got until you open it up. Chal-

A demonstration at a gem show. Cracking open a geode requires specialized equipment—and maybe some showmanship.

cedony and agate are about a 7 on the Mohs scale of hardness—hard, but not *too* hard. This makes the insides of geodes, thundereggs, and agates the perfect material for rock hobbyists and jewelers.

First, the geodes are cut into thin or thick slabs. The next part is the hardest—deciding how to bring out the natural beauty of the stone. From bookends and clocks to ornate carved cameos and rings, the journey begins with recognition and admiration of that natural beauty. Once the choice is made, the artist begins. The slabs can

GEODES

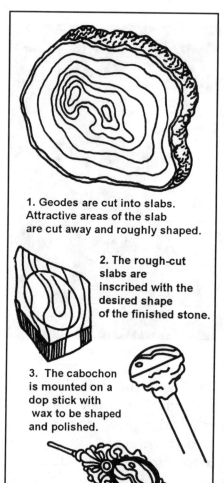

1. Geodes are cut into slabs. Attractive areas of the slab are cut away and roughly shaped.

2. The rough-cut slabs are inscribed with the desired shape of the finished stone.

3. The cabochon is mounted on a dop stick with wax to be shaped and polished.

4. The highly polished stone is mounted in a setting.

be cut and faceted into cabochons, or "cabs," the dome-shaped stones that are mounted into rings, bolo ties, or belt buckles. Cabs come in many sizes and shapes, from ovals and rectangles to squares and triangles.

The larger slabs can be polished and used for paperweights, bookends, or clock faces. The uses of the geode's durable crystal are limited only by the imagination.

Perhaps you can find or discover a formation that brings fame to your city or state. That's exactly what happened in Oregon, where the state rock is the thunderegg. The Cascade Range south of Portland is home to many deposits of these colorful nodules. The local name thunderegg comes from the legends of the Native Americans of

In Search of Geodes

Agate geodes can be left in their natural state, or they can be made into decorative bookends and other useful items.

central Oregon. Rock hounds the world over have adopted the name thunderegg from this myth.

Eggs down under, anyone? Thunderbird Park on

The Badlands of South Dakota are rich in geode treasures.

In Search of Geodes

Tamborine Mountain in Australia is proud to bear that name. Located in the crater of a dormant, or inactive, volcano is a camp where you can hunt for eggs all year. These are not regular eggs; some of them have their own supply of oil, not to mention some of the world's most colorful chalcedony, left behind by a giant thunderbird, according to local legend.

While you are playing in the surf in Florida, you might discover some interesting coral formations. Some of the coral around the Tampa Bay area is filled with agate. Much like a geode which contains water, these coral branches have water sealed inside.

The Badlands of the Dakotas have had their share of bad press, but when it comes to agates and geodes, they're the tops. The Fairburn agates from the Black Hills of western South Dakota are world-famous for their shape and color. A good Fairburn is full of intense banded colors or filled with patterns unique to this badlands' location. South Dakota is a gem-hunter's paradise: Petrified Wood Park, Jewel Cave National Parkland, and Homestake Gold Mine are but a few sites on the must-see list for rock hunters. Keep in mind, though, that taking rocks from national parks is not allowed!

Our neighbors south of the border also offer some of the most abundant geode formations ever found. Chihuahua, Mexico, is a leader in agates and geodes of all colors and sizes. The ranches of Northern Mexico, such as the Laguna Ranch, are famous for the colorful agates found there.

GEODES

Further south, the hills and mountains of Brazil have yielded some of the largest geodes ever found. The record may be held by the 412-pound, amethyst-lined geode mined from the Rio Grande de Sul. This incredible specimen is deep purple and large enough to fit over your head.

Closer to home, southern Utah is home to some of the most colorful sedimentary geodes ever found. Formed

The forces of erosion created this bridge at Arches National Park, Utah. Utah's geodes formed under ancient oceans.

In Search of Geodes

long ago at the bottom of oceans, the Septarian Geodes are brilliant examples of crystalline structure. Yellow and orange geometric patterns are the trademark of these nodules.

The colors and shapes found in each geode are often clues to its origin. There are purple agates from Arizona and Chihuahua and Durango, Mexico, and red and orange agates from Yellowstone River in Montana; Burro

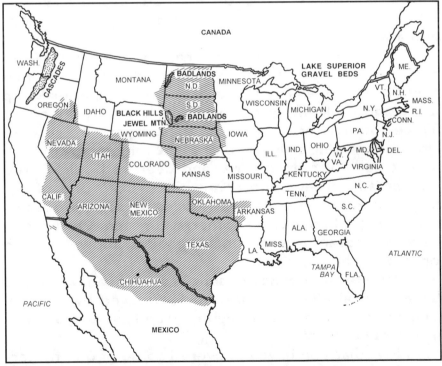

The best places to look for geodes are in the desert areas of the Southwest (shaded portions of map). But if you look high and low, you may find geodes around the Great Lakes, and in gravel beds, dry lake beds, and at shorelines.

Creek, Arizona; Fairburn, South Dakota; and Chihuahua, Mexico. Brilliant blues, on the other hand, could mean they come from Ellensburg, Washington, while deep greens possibly indicate a source in Montana.

Colors and Minerals

The color in geodes is easy to explain. The mineral layers separate themselves in some pre-defined order as they interact. Crystallization, magnetism, and gravity all play a role in the familiar banding of geodes, thundereggs, and agates.

The palette of colors inside each nodule comes from the varied kinds of minerals present during the formation of the stone. The tiny quartz crystals that make up agate and chalcedony are originally a soft, milky color. Traces of minerals that invade these layers are responsible for creating brilliant colors. The mineral manganese adds traces of purple, pink, and deep red. Nickel and copper add greens and some blues. Iron, one of the most common minerals, is the easiest to spot, producing red, orange, brown, and sometimes green. Quartz and calcium, present in sedimentary geodes, add yellow and orange to the crystals.

Some of the colors present in agates can be caused by an optical effect like the one produced by a prism. The crystals separate enough to trap light and break it into a rainbow of colors.

Now that you have uncovered some of the mysteries of the geodes, get out there and hunt some up!

In Search of Geodes

Rings of agate in a Laguna geode found in Chihuahua, Mexico.

FOUR

How to Make Your Own Natural-Looking Geodes

You will need the following things to grow your twin "ruby" geodes:

> *Another Favorite From Mother Earth's Kitchen*
>
> **A Recipe For:** TWIN "RUBY" GEODES
> 40 GRAMS OF GYPSUM (INCLUDED IN THIS KIT)
> 40 GRAMS OF ALUM (INCLUDED IN THIS KIT)
> 150 GRAMS OF ALUM, COLORED RED (INCLUDED IN THIS KIT)
> 1 GEODE MOLD (INCLUDED IN THIS KIT)
> 1 SPATULA (INCLUDED IN THIS KIT)
> 2 TEASPOONS WATER
> 1 HEAT-RESISTANT GLASS OR PLASTIC CONTAINER MEASURING AT LEAST 5 INCHES SQUARE AND 6 INCHES HIGH (A PLASTIC GALLON JUG WITH TOP CUT OFF WORKS FINE.) THIS IS YOUR CRYSTAL GROWTH CONTAINER.

Making the Geode Shells:

1. Slit open the bag containing 40 grams of white gypsum and the bag containing 40 grams of alum (single white crystal). Pour the contents of both bags on a sheet of newspaper and mix thoroughly with spatula. Divide mixture into two equal amounts.

2. Pour half of mixture into the geode mold.

3. Pour 2 teaspoons water into mold.

4. Stir mixture and water with the spatula. As mixture thickens, spread evenly on the inner wall of the mold up to the rim. Mixture hardens quickly! Scrape along the top rim of the mold to make sure the edge is smooth. This will form the geode shell.

5. Let the geode shell harden for 30 minutes.

6. Pop the shell out of the mold, then repeat steps 2 through 5 with the remaining mixture to make a second geode shell.

Growing Ruby-Colored Geodes

1. Slit open the bag containing 150 grams of dyed red alum and pour contents into the growth container.

2. Add 22 ounces of boiling water. Stir gently until the chemical is dissolved. Then let the solution cool for 20 minutes.

3. Using rubber gloves, carefully place geode shells into the solution with the openings facing sideways. Make sure they don't touch each other, and that they are completely covered by the solution. Place the growth container in a room with a constant temperature, and where you can check on it without disturbing it. You can use a flashlight to check the crystal growth occasionally, but be sure not to tip over the geode shells.

4. Let the crystals grow for about one week. Then pour the solution down the drain while the water is running. Wearing rubber gloves, remove the geodes. Put them on newspaper to let them dry thoroughly.